LOST IN SPACE AND TIME

Lisa Visintainer

Ilias Thiesseas
6774 Tschagguns, Austria

First edition

The original work was written in German and published under the title
Zwischen Raum und Zeit
Translated by Ilias Thiesseas

Printed in Poland

Front cover image by Ilias Thiesseas
Book design by Ilias Thiesseas

Book ISBN: 9798644325658

ABOUT THE PUBLISHER

Ilias Thiesseas is a professional publishing company founded in Austria, following a clear goal: To benefit society and authors, giving valuable information to the world.

We, at *Ilias Thiesseas*, created a new branch in our company that supports new authors.

Note that *Lost in Space and Time* by Lisa Visintainer is a part of our *VWA Publikation Professional Publishing Program™*, an increasingly popular program for first-time writers to get their ideas into the world. We want authors to focus on what's important to them: writing. Ilias Thiesseas deals with the publishing and also contributes to junior writers' financial independence, so they can fulfil their dreams.

ABSTRACT

When it comes to time travel, even among scientists there are controversial discussions. While
some natural scientists are firmly convinced that
travel through time is quite possible, others reject
it completely. Therefore, it would be interesting to
know whether time travel is science or just fiction.

As an introduction, the work deals with Albert Einstein's theories of relativity, which today form one
of the most important foundations of physics. They
explain many aspects of the universe and thus also
play an important role in time travel. After this
introduction, black holes are briefly discussed, as
they could play an important role in time travel.
Then the concept of time is examined in more
detail. The present paper should show how time
is defined or which different approaches are used
to describe time, for example by so-called time
arrows. Furthermore, time machines or methods
with which it would be possible to travel through

time, such as a "special relativistic turbo capsule", a "general relativistic parking capsule" or wormholes, are analyzed. The work is concluded with the paradoxes that may arise under certain circumstances, such as the grandfather paradox or the twin paradox, and their solutions.

For the work, mainly book material from renowned physicists has been used. However, established magazines and online media have also been chosen for more generally known information or current events.

CONTENT

9. Conclusion

References

List of figures

If time travel is possible, where are the tourists from the future?

– STEPHEN HAWKING (CF.: KAKU, 2016, P. 276)

'[Time traveling] contradicts logic,' said Filby. 'What logic?' said the time traveler.

– H. G. WELLS (CF.: KAKU, 2016, P. 276)

People who believe in physics, like us, know that the distinction between past, present and future is only a particularly persistent illusion.

– ALBERT EINSTEIN (CF.: KAKU, 2017, P. 367)

"Scientists know very well that time is only a type of space"

-(CF.: MÜLLER, 2016, S. 19)

1. INTRODUCTION

These are just some of the aphorisms I came across while writing my book. They show how different opinions are about time, space and time travel. The one or other aphorism is also the reason for the creation of this book. Especially that one by Stephen Hawking has kept me very busy. Because he is probably right: Nobody has ever met a time traveler from the future. If time travel were possible, one or the other time traveler could surely have strayed in our present.

The universe is very multifaceted. It includes the world of the smallest particles to the largest suns and galaxies. Since my childhood I have read books about our solar system, the structure of planets and stars. My interest started with the book "Your colorful dictionary of space", which explained some aspects of the universe in simple language and with illustrations. The universe has fascinated me all my life. Many hobbies come and go, and as you get older, your interests often change. It has been the same with me. However, the mysteries of the universe have been with me since I was a child and that is why I decided to formulate a problem in connection with the universe for my book.

I came to the specified field of time travel primarily through the film trilogy "Back to the Future" from the 1980s. I liked the combination of humor and science very much, and I wanted to learn more about time travel and if this is really possible. The film "Interstellar", which was released in 2014, also deals with time travel. The theoretical physicist Kip Thorne was involved in the creation of the film. Thus, most of the physical aspects in the film are scientifically based. So, it is not only (science-) fiction (cf.: Hummel, https://www.zeit.de/wissen/2014-11/interstellar-physik [accessed: 06.02.2020]).

I also read about an "experiment" by Stephen Hawking, in which he wanted to show that time travel is not possible. He had organized a party for time travelers in 2009. However, nobody came to the celebration, because he had sent the invitations the day after (cf.: Hawking, 2018, p. 168).

If it were possible to travel through time, it would change humanity. What would happen if everyone had the possibility to move back and forth in time? Finding out if it is possible or not is therefore of great importance. This given the fact that we have never had so many technical possibilities as today.

The literature used for the book consists mainly of print media. The sources are from high-ranking physicists who have also been very interested in time in general. I read the definitions and opinions about the different problems and compared them to draw my own conclusions from many different views.

In my book I will try to find out if time travel is possible or if it belongs more to the science fiction scene. At first, I will give a compact description of the most important results of Albert Einstein's theories of relativity (special and general), because today they are the basis for the explanation of all physical events in the universe and thus basically also for time travel. In addition, I will deal with the concept of time, and the construction of time machines will be discussed. Further on I will discuss possible time paradoxes and the solutions proposed for them.

2. THE SPECIAL THEORY OF RELATIVITY (STR)

The probably most famous physicist of the 20th century, Albert Einstein, completely changed the world view in physics with his two theories of relativity. The STR, which was published in 1905, has become one of the most important theories of the 20th century (cf.: Apolin, 2008, p. 4). Until 1905, space and time were seen as absolute and unchangeable. Einstein's STR and General Theory of Relativity (GTR) describe not only processes in the universe but also events on Earth. However, in everyday life these effects are not noticeable to people, as they are very small and hardly measurable (cf.: Putz/Jahn, 2019, p. 201).

There are two different types of movements. At the one hand, there are unaccelerated (uniform) and at the other hand accelerated (non-uniform) movements. The STR deals exclusively with un-accelerated systems, i.e. movements that occur in a straight line and with constant speed (cf.: Apolin, 2008, p. 6). Then the first Newtonian axiom, also known as inertia theorem, applies. Systems in which the law of inertia applies are also called inertial systems (cf.: Putz/Jahn, 2019, p.203).

Einstein formulated two postulates that form the basis of STR:

> 1. *There is no excellent inertial system All inertial systems are equal and physical processes take place in them in the same way.*
>
> 2. *The vacuum light speed is constant. It has the same value for all observers regardless of whether the light source or the observer (or both) is moving". (cf.: Putz/ Jahn, 2019, p. 205).*

From the above postulates, Einstein concluded that time, from the point of view of a resting observer, passes more slowly for an observer in motion. This is also called time dilation.

For the time dilation applies:

$$t_b = t_r \cdot \sqrt{1 - \frac{v^2}{c^2}}$$

Here t_b stands for the time that elapses for the observer in motion, t_r for the time that elapses for the resting observer, **c** for the speed of light and **v** for the relative speed (cf.: Apolin, 2008, p. 16 f.)

There are also consequences for the spatial expansion of moving objects at higher speeds. Objects that move relative to an observer are contracted in the direction of movement (cf.: Putz, Jahn, 2019, p. 211).

The following applies to the length contraction:

$$l_b = l_r \cdot \sqrt{1 - \frac{v^2}{c^2}}$$

Here l_b stands for the length measured by the observer in motion, l_r for the length measured by the observer at rest, **c** for the speed of light and **v** for the relative speed. (cf.: Apolin, 2008, p. 21)

Not only time and lengths in direction of movement, but also mass of bodies depends on speed. Einstein could deduce following relation within the STR:

$$m_D = \frac{m_0}{\sqrt{1 - \frac{v^2}{c^2}}}$$

Here m_D stands for the dynamic mass, m_0 is the rest mass, **v** the relative velocity and **c** the velocity of light.

However, the relativistic increase in mass does not mean that a body expands, but that its inertial mass (cf.: Apolin, 2008, pp. 25-26), which indicates the acceleration with which an object reacts to an acting force, increases (cf. Pössel, https://www.einstein-online.info/spotlight/traegeschwere/ [accessed: 05.02.2020]).

Another important conclusion from the STR is the relationship between energy and mass. It applies:

Total energy = rest energy + kinetic energy

The result eventually sums um to: $\mathbf{E=mc^2}$

This famous formula is known as the equivalence of mass and energy (cf.: Apolin, 2008, p. 27).

3. THE GENERAL THEORY OF RELATIVITY (GTR)

Since STR refers exclusively to inertial systems, Albert Einstein wanted to develop a theory that would apply to accelerated systems (cf.: Putz/Jahn, 2019, p.231).

In the GTR, Einstein introduced the equivalence principle:

Inert and heavy masses are always the same size because they are the same and therefore indistinguishable (cf.: Apolin, 2008, p. 39).

It does not matter whether an astronaut accelerates a rocket with 9.81 m/s² or stands on the earth in mid-latitudes. Similarly, it cannot be distinguished whether the rocket is floating at a very great distance from a celestial body or whether it is in free fall. (see: Apolin, 2008, p. 39). From these findings, Einstein formulated an extended form of the equivalence principle:

By no experiment whatsoever, can one distinguish between inertia and gravity. (cf.: Apolin, 2008, p. 40)

It also follows from the equivalence principle that light in gravitational fields must be deflected. A photon, a particle of light flying through an accelerating rocket, flies horizontally through the rocket from the outside, as seen by an observer. For an observer inside the rocket, however, the photon has a parabolic path (cf.: Apolin, 2008, p.40).

Furthermore, the GTR states that gravity also has an influence on the course of clocks. If an observer A sends a photon vertically upwards in a gravitational field, it must lose energy due to the gravitational force. The wavelength of the photon increases and its frequency decreases. If now another observer B sends a photon down, i.e. to the surface of a celestial body, it receives more energy and thus a higher frequency. Observer A senses that a watch worn by observer B goes faster. For B, the clock that A carries goes slower, because A is closer to the mass.

The time change in an inhomogeneous field of a large mass is described by the following formula:

$$T_A = T_B \cdot (1 - \frac{G \cdot M}{c^2 \cdot r})$$

In this formula T_A is the time that elapses for an observer who is far away from a mass, for example the earth's surface, and T_B is the time that elapses for an observer who is near a mass. **G** stands for the gravitational constant G = $6{,}67 \cdot 10^{-11}$ Nm²/kg², **M** is the mass in kilograms, **c** is again the speed of light and **r** stands for the radius of the central mass in meters (cf.: Apolin, 2008, p. 43 f.).

Gravity therefore has an influence on the movement of clocks, which run slower when they are near massive objects.

However, since the speed of light is always about 300,000 kilometers per second, scales in the gravitational field must shrink by the same factor by which time is also slowed down. Thus, the GTR also states that scales shrink in a gravitational field.

For a change in length in an inhomogeneous field of a large mass results in the following relationship:

$$L_A = L_B \cdot (1 - \frac{G \cdot M}{c^2 \cdot r})$$

This formula is basically the same as that of the

time change in an inhomogeneous field of a large mass. In this formula, however, L_A stands for the length of a scale in the vicinity and L_B for the length of a scale in the far distance of a mass (cf.: Apolin, 2008, p. 45).

4. BLACK HOLES

B lack holes play an important role in some theories about time travel. These will be discussed in more detail in the course of this book. The following information should explain how these objects are created, how they are constructed and how they can be detected.

Stars become red giants in their final stages and, depending on their mass, they develop into white dwarfs, neutron stars or black holes. When the mass of a star is several solar masses, it collapses in on itself to a point where its density is infinite. Within the Schwarzschild radius, named after the German astronomer Karl Schwarzschild, nothing can escape, not even light, because of its strong gravitation. This is why these objects are called black holes. (cf.: Apolin, 2008, p. 50).

To detect black holes, the kinematic method is a promising approach. Here, the orbits of visible bodies are examined and then the properties of invisible bodies are deduced. They could also be indirectly detected with the help of so-called accretion disks that form around the black hole. These disks emit highly energetic radiation (cf.: Apolin, 2008, p. 51). Another possibility is the so-called Hawking radiation, named after the English physicist Stephen Hawking. He postulated that black holes need not exist forever. If a pair of a positron and an electron forms near the Schwarzschild radius of a black hole, one of the particles could fall into the black hole and the other escape. As a result, its mass would have to be reduced by the value of the mass of the particle that escaped until the black hole is completely dissolved (cf.: Jaros/Nussbaumer/ Nussbaumer/ Kunze, 2007, p. 98). Another finding from the GTR is that the denser an object is, the more space-time bends in the vicinity of the object. The curvature of space-time for a black hole becomes infinitely large. One can imagine this phenomenon as a funnel with vertical walls (see Fig. 1) (cf.: Apolin, 2008, p. 51).

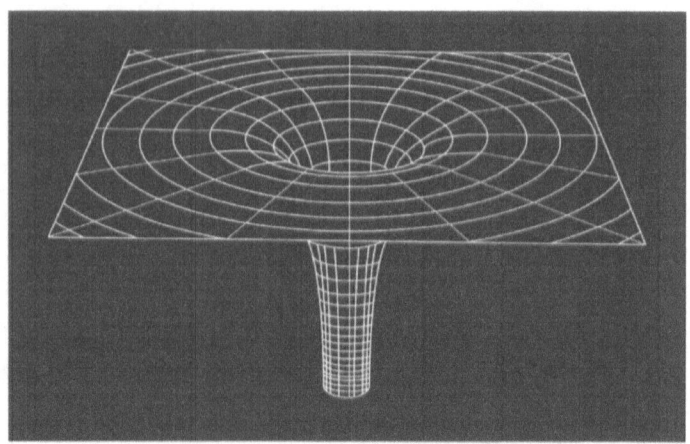

Figure 1: Curvature of black holes

In April 2017, scientists used eight radio telescopes located at different positions around the world to photograph the black hole at the center of galaxy M87. The resulting false-color image (see Fig. 2) naturally does not show the black hole itself, but only its surroundings (cf.: Gast, https://www.s-pektrum.de/magazin/ins-herz-der-finsternis-das-erste-bild-eines-schwarzen-lochs/1647844 [accessed: 06.09.2019]).

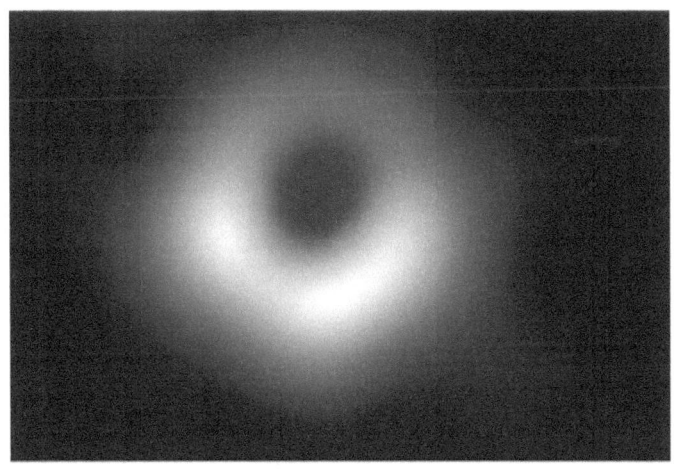

Figure 2: The first image of a black hole

5. TIME

Time seems to have no influence in everyday life, because you cannot simply stop or accelerate it. Since it passes uniformly and you cannot influence it, this time is also called absolute time. Time is not a substance (not matter), but it is measured by clocks, which are material objects. For people on earth, time, together with space, is a decisive factor, although space and time are very different (cf.: Müller, 2016, p. 2).

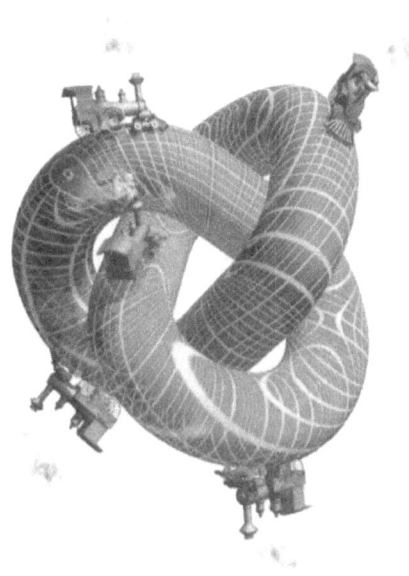

Figure 3: Form and direction of time

Fig. 3 shows the relationship between space and time as predicted by Einstein's theory of relativity. Space cannot be curved if time is not also taken into account. Time therefore also has a kind of shape, like space. However, the locomotives in the figure show that time has only one direction, because they only travel in one direction (cf.: Hawking, 2016, p.41).

5.1 The time arrow

Space consists of three dimensions in which every-thing can move in all directions. Time, on the other hand, moves in only one direction: the future. This property of time is also called time arrow. To understand this, a parameter in the second law of thermodynamics (thermodynamics) helps. This law states that entropy, the measure of disorder in a system that is complete, never decreases. It can only either remain the same or above all increase. Entropy can be described with the help of a simple thought experiment from everyday life (cf.: Müller, 2016, p. 2 f.). When a cup stands on a table, it is in a state of high order, i.e. it has a low entropy (cf.: Müller, 2016, p. 3 f.). If now the cup falls off the table and breaks on the floor, its state has changed from a high order to a high disorder. The entropy of the cup has thus increased. But if the broken cup should now become a whole again, this does not work so easily. It can be glued back together again, but this requires energy, and so in the end, because of the use of energy, the result would still show an increase in entropy (cf.: Müller, 2016, p. 4).

Since this method of calculating the direction of time with the help of thermodynamics can be justified, this resulting time arrow is also called a thermodynamic time arrow. In relation to the universe, entropy is also steadily increasing and the term thermodynamic time arrow is transformed into the term cosmological time arrow. However, it is not yet certain whether our universe is really a complete system (cf.: Müller, 2016, p. 5).

There are at least three time arrows in total. The

thermodynamic and cosmological arrow just mentioned and also the psychological arrow. Fig. 4 shows the three mentioned
Time arrows (cf.: Hawking, 2019, p. 187).

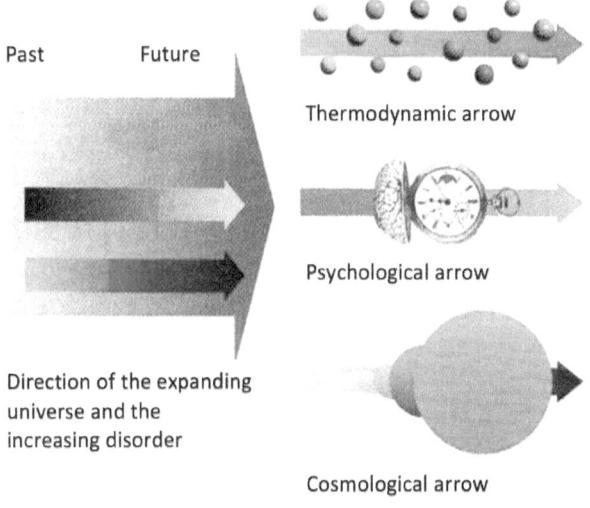

Figure 4: The time arrows

The psychological time arrow can be equated with the thermodynamic arrow. The psychological time arrow stands for the passing of time as we as humans perceive it. It also describes that we can only remember the past and not the future (cf.: Hawking, 2018, p.187 f.). With the help of computers, the psychological arrow can be explained, because it should be the same for computers as for a human being. A memory of a computer consists of building blocks that can take on two states. This is comparable with an abacus. Here single balls can only be moved either to the right or the left side. Before some thing is stored on a computer, this information is in a state of disorder (cf.: Hawking, 2018, p. 189). When the computer memory interacts with the system to be stored, it assumes one state or the other, like the abacus balls. The computer's storage system has thus been transformed into an ordered state. In order to finally bring the memory into the right state, energy must be expended. For the abacus, you have to move a ball, you have to supply a computer with electricity. This supplied energy is transformed into heat, which is given off and the disorder in the system is increased. The time direction for entropy is the same as the time direction of a human memory, because we remember events in a certain (ascending) order, just as entropy increases. It follows that the thermodynamic time arrow determines our sense of the time direction (i.e. the psychological time arrow).

The cosmological time arrow already mentioned above stands for the expansion of the universe (cf.: Hawking, 2018, p. 190). Since the universe originated from a big bang singularity, it cannot be said with absolute certainty how it began, because the laws of nature lose their validity with singularities. However, it is likely that it began in a state of low entropy. This would also indicate why a precisely defined thermodynamic time arrow can be observed today. If, on the other hand, the universe had begun in an already disordered state, entropy could hardly increase any further. This would mean that no thermodynamic time arrow could be defined. If the entropy were to decrease, the cosmological arrow would point in the opposite direction of the thermodynamic arrow. (cf.: Hawking, 2018, p. 191).

5.2 Spacetime

Another achievement of Albert Einstein is the discovery of the existing relationship between space and time. In 1908, the German physicist Hermann Minkowski already postulated this unity in a lecture. He said: "From 'hour' on, space for itself and time for itself shall sink completely into shadows and only a kind of union of the two shall preserve independence". (cf.: Müller, 2016, p. 8). He called this union space-time-continuum or rather known as space-time. Einstein's GTR describes that we live in a dynamic, curved and four-dimensional space-time. In the GTR a space-time is spanned by four dimensions, namely height, length, width and time. In the STR it is a flat spacetime, i.e. a two-dimensional plane like a chessboard. In contrast, in the GTR spacetime is curved by masses (cf.: Müller, 2016 p. 12).

Every celestial body has its own spacetime and the whole universe can be described by a single spacetime, which is called Friedmann universe (cf.: Müller, 2016, p. 9). These spacetimes can be calculated by complex mathematical equations from Albert Einstein's GTR, called Einstein's field equations. Solutions of these equations published in November 1915 are spacetimes. The first and most important solution of the field equations, the so-called Schwarzschild solution, was found by Karl Schwarzschild in 1916 and is used to describe the gravitation of planets, stars and electrically neutral non-rotating black holes. (cf.: Müller, 2016, p. 10).

6. TIME
MACHINES

I n the past, many suggestions have been made by physicists as to how a machine could look and function that would allow people to travel through time. Some ideas were rejected because they contradicted Einstein's equations and could not be physically proven. However, there are also some concepts that could possibly be put into practice sometime (cf.: Kaku, 2017, p.387).

6.1 Special-relativistic turbo capsule

This concept of a time machine works by means of very high speeds. We're moving here at the speed of light. Among other things, it is based on the principle of the special relativistic time expansion from the STR.

If you want to accelerate a space capsule to speeds close to the speed of light, this proves to be almost impossible, because the mass of the capsule would be infinitely large and the time dilation effect would increase immensely. Assuming this would be technically possible, a time expansion factor of about ten could be assumed. If it were possible to maintain the relative speed, which would be close to the speed of light, of a capsule for one hour, a time traveler in the space capsule would age ten times less.

On Earth, 60 minutes would pass while for the time traveler, just six minutes would pass. The time traveler would therefore have travelled 54 minutes into the future.

Accelerating a spaceship to such high speeds, however, brings with it some problems in the areas of technology, energy and safety (cf.: Müller, 2016, p. 34). With regard to the energy input, the particle accelerator LHC (Large Hadron Collider) at CERN in Geneva can be taken as an example. In a ring-shaped particle accelerator, lead ions or protons are brought to relativistic speeds. This generates magnetic fields with a field strength of 8 Tesla (cf.: Müller, 2016, p. 34 f.). For macroscopic objects such as a space capsule with a human being inside it, such methods cannot yet be used today because the energy input would be far too high. Such high-energy experiments are a very big step in the right direction in terms of time travel.

The highest speed at which macroscopic objects such as space probes have been accelerated so far is just above the escape velocity of the Earth, which is about 40,000 km/h (cf.: Müller, 2016, p. 37). To fly faster in space with a space probe there is the so-called swing-by-maneuver. Here a space probe is steered towards a planet and accelerated by its gravity.

If you want to accelerate a capsule, not only the mass but especially the speed causes a big problem, because if a body moves with double speed, you need four times as much kinetic energy to bring it to higher speeds (cf.: Müller, 2016, p. 38). In the case of masses, the law of inertia, according to which all bodies that are in motion remain in motion and all those that are at rest wish to remain at rest, is responsible for the difficulty of acceleration. The greater a mass is, the greater the inertia. In addition, the faster a mass is, the more difficult it is to accelerate it even more.

Furthermore, traveling at high speeds always brings with its safety risks. The faster an object moves, the more difficult it is to steer it safely. This effect can also be felt when driving a car, for example when driving at almost 200 km/h. An object that moves at almost the speed of light is naturally very difficult to maneuver (cf.: Müller, 2016, p. 39). That's why it's better to accelerate a space capsule for such an undertaking in a circular fashion, like in a particle accelerator, and not in a linear fashion. But the space capsule would become faster and faster, the centrifugal force would increase and it would become more and more difficult to keep the capsule on track. Nor should one think about where one is flying to, because at relativistic speeds, thinking about whether to turn right or left could end fatally.

Moreover, at such high speeds, even the smallest microparticles, which are abundant in interstellar space, can become a major problem (cf.: Müller, 2016, p. 40). The particles have high kinetic energies and can therefore cause great damage and destroy the turbo capsule.

There is also the problem of electromagnetic radiation, which can be very dangerous at relativistic speeds. The reason for this is the Doppler effect. An astronaut in a spacecraft perceives the oncoming radiation more intensively and at shorter wavelengths. At relativistic speeds, the ambient radiation also shifts, and the space capsule with the astronaut would thus be pierced by the highly dangerous gamma radiation (cf.: Müller, 2016, p. 41).

Figure 5 shows how the environment changes during relativistic movements. The faster one moves, the more the space traveler can observe what is opposite to his direction of movement (Fig. 5 b). This also explains why the radiation becomes more intense from the front, as all light particles hit the front of the spacecraft. The effect that results is called the relativistic light aberration or beaming effect (cf.: Müller, 2016, p. 43 f.).

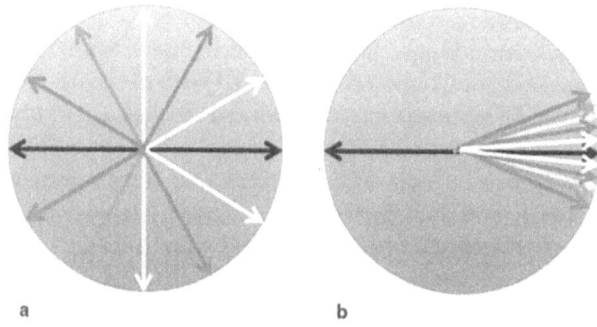

Figure 5: Doppler effect at relativistic velocities

Flying through space with a turbo capsule and thus travelling into the future is therefore proving to be an impossible challenge today. The technology, safety and control of the capsule do not make it possible to build or use such a time machine. (cf.: Müller, 2016, p. 43).

6.2 General-relativistic parking capsule

This concept for a time machine is based on the principle of time dilation from the GTR, in which time passes more slowly near large masses. A time traveler would travel to an object with a large mass, e.g. a black hole with a corresponding flying machine and stay there for a longer time on its surface or in the area of the event horizon. If the time traveler then moves away from the object again, he has travelled into the future, since he then returns to those areas where time has passed more quickly (cf.: Müller, 2016, p. 49).

6.2.1 Time travel using various objects in the universe

The table below shows the time dilation on the surface of different bodies. The last column shows the elapse of time on the surface of the objects while 60 seconds elapse in a reference system which is infinitely far away from the object (cf.: Müller, 2016, p. 49f). As can be seen from the table, the mass of the Earth is not suitable for time travel, because while 60 seconds pass in the reference system, 59.99999996 seconds pass on the Earth. A time traveler would therefore travel into the future by just 0.00000004 seconds. In contrast, time stops at the event horizon, the surface, of a black hole. So to achieve a reasonable effect, you need at least one neutron star or - of course the most effective method - a black hole.

"

Object	Mass (kg)	Radius of the surface (km)	Gravitational time dilation effect	Time interval Δt
Earth	$6 \cdot 10^{24}$	6367	≈ 1	59,99999996
Jupiter	$2 \cdot 10^{27}$	70.000	0,99999998	59,9999988
Sun	$2 \cdot 10^{30}$	700.000	0,999998	59,99987
White Dwarf	$2 \cdot 10^{30}$	5000	0,9997	59,9823
Neutron Star	$4 \cdot 10^{30}$	10	0,64	38,39
Black Hole	$10 \cdot 10^{30}$	15	0	0

" (cf.: Müller, 2016, S. 50 f.)

6.2.2 Time travel using a neutron star or a black hole

Assuming a time traveler spends a year on the surface of a neutron star, he would then travel one year into the future, since twice as much time has passed at a distance from the neutron star. But of course, this causes enormous technical, health and time problems. The nearest neutron stars are some hundred light years away. So even with advanced rocket technology you would have to travel a very long time to reach a neutron star at all. If you were to travel with your time machine at relativistic speeds, you would again have the problem of interstellar microparticles and dangerous radiation, which is blue-shifted due to the Doppler effect, as it moves towards the space capsule. In the 1970s, the British Interplanetary Society's Daedalus project aimed to create a suitable space capsule for such interstellar travel. The project was never realized, since both the technical and financial hurdles seemed insurmountable (cf.: Müller, 2016, p. 51f).

However, if one now assumes that a spaceship could reach a neutron star, the problem of landing would arise. Because the gravitational forces are so enormous, it would be extremely difficult to land gently on the surface. On a neutron star, the gravitational acceleration compared to the Earth is about 100 billion g, while g is the gravitational acceleration of the Earth, which is about $10 m/s^2$. A crash could therefore not be avoided with the current state of the art (cf.: Müller, 2016, p. 52).

If these technical hurdles could be overcome, a time traveler on the surface of a neutron star would again be exposed to very dangerous radiation.

On the one hand, radiation that is generated around the neutron star, and on the other hand, radiation from the farther surroundings reaches the neutron star. An additional problem would be that an astronaut would literally be crushed by the immensely strong gravitational forces.

Assuming that one would survive all these strains and manage to land on the surface and "park" there for some time, how would one manage to leave the neutron star again? The escape velocity, which in the case of a neutron star is about three quarters of the speed of light, would have to be immensely high again, as it was when the neutron star landed (cf.: Müller, 2016, p. 53 f.).

For a time traveler, black holes would cause the same problems as described above, but on these objects they would be even more extreme. Above all, "parking" would prove to be particularly difficult, because black holes, unlike neutron stars, do not have a solid surface (cf.: Müller, 2016, p. 54).

Such a "parking" on an object with a large mass in order to travel into the future is therefore theoretically possible in thought experiments, but we are still very far from a practical feasibility (cf.: Müller, 2016, p. 57).

6.3 Wormholes

Wormholes are constructs of theoretical physics that, like a tunnel, represent a connection in space-time. With their help one could travel through space and possibly simultaneously through time (cf.: Vaas, p. 167 f.). The name is somewhat deceptive, because a wormhole is not empty like a hole, but consists of space (cf.: Vaas, 2013, p. 178). In 1935, Einstein published an article together with the American physicist Nathan Rosen, which proves that the GTR allows wormholes, then called "bridges" (cf.: Hawking, 2018, p. 205). Its name was coined by the American physicist John Archibald Wheeler. He called the objects wormholes because he compared them to a worm crawling through an apple. The path or channel through which the worm bites its way is similar to the idea of a wormhole (cf.: Vaas, 2013, p. 166).

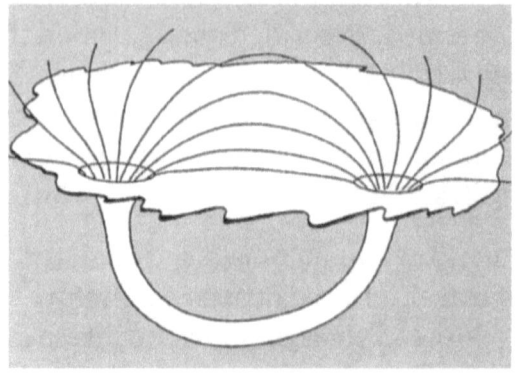

Figure 6: The first drawing of a wormhole

Fig. 6 is the first draft of a wormhole that Wheeler created in 1955 (cf.: Vaas, 2013, p. 168). Wheeler assumed that wormholes would give space-time a structure that is "foamy". However, this happens only on very small scales. Thus, Wheeler introduced the concept of a so-called space-time foam, in which wormholes are a component (cf.: Vaas, 2013, p. 169). Although they only exist on paper so far, the American physicist Kip Thorne found out that Einstein's theory of relativity would definitely allow wormholes and thus "faster-than-light" journeys. (cf.: Vaas, 2013, p. 166).

In a wormhole, a black hole could form the entrance and a so-called white hole the exit. A white hole is the antagonist to a black hole - so everything flies out of it. By reversing the time direction in the GTR, a solution of the field equations is obtained that points to white holes (cf.: Müller, 2016, p. 58). White holes cannot absorb matter, but only emit matter. That's why some physicists also suspect that the Big Bang was a white hole. However, the existence of these objects could not yet be proven, as they would be very unstable and would very quickly transform into black holes (cf.: Vaas, 2013, p. 52).

6.3.1 Requirements for a wormhole

In order to travel through space and time with a wormhole, some important prerequisites would be necessary.It must not move, it must be stable and an event horizon must not enclose it, otherwise you would be trapped in the wormhole. Furthermore, the gravitational forces in the wormhole must be small, otherwise the traveler would be torn apart (cf.: Vaas, 2013, p. 172). Finally, energy and matter should not be required in "infinite" quantities. Kip Thorne and Michael Morris actually found a solu-tion to the Einstein equations that does not exclude the existence of wormholes. The geometry of this solution can be imagined similar to an hourglass (cf.: Vaas, 2013, p. 173.).

6.3.2 The construction of a wormhole

To travel through a wormhole, you must first dis-cover or construct one. One could, for example, en-large a wormhole hidden in the space-time foam so that, for example, a human being or another macroscopic body would fit through it. Another way to "build" such an object would be to deform spacetime, cut it open and connect the resulting open edges.

The greatest difficulty in using a wormhole as a time machine would be to keep the opening, called the gullet, stable and open. Wormholes would be very unstable objects, but immediately after the Big Bang they could have been stabilized by so-called Cosmic Strings with the help of negative mass and still exist. (cf.: Vaas, 2013, p. 175 f.). Cosmic strings are very long threads that have an extremely small cross-section and could travel at almost the speed of light due to their high tension. They could have formed only a short time after the Big Bang (cf.: Hawking, 2018, p. 157).

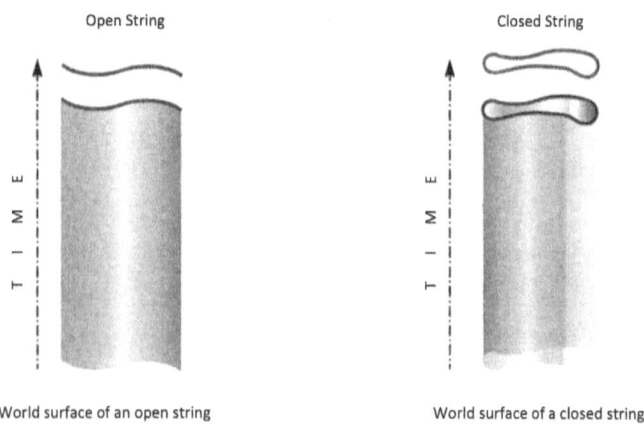

Figure 7: Strings and their world surface

In fig. 7 you can see an open string with its world area on the left. On the right, a closed string and its world surface is shown, which forms a kind of tube or cylinder (cf.: Hawking, 2019, p. 217).

In order to keep a wormhole open, a very special material is needed - so-called exotic matter, which has a negative mass or energy density, because this can cause a gravitational repulsion in the tunnel (cf.: Vaas, 2013, p. 176). Fig. 8 shows the difference between ordinary and exotic matter. The surface of a sphere (such as the Earth) is curved outwards, so it is ordinary matter as shown in Fig. 8 below. However, a wormhole must curve inwards. Fig. 8 below shows the curvature caused by exotic matter. This negative curvature resembles a saddle. (cf.: Hawking, 2019, p. 204).

Figure 8: Positive and negative curvature of surfaces

To prevent the wormhole from collapsing or being torn apart by a passing spacecraft (see Fig. 9), the exotic matter is needed to build up negative pressure and keep the wormhole stable (cf.: Vaas, 2013, p. 176).

Figure 9: Einstein-Rosen Bridge

The exotic matter can be proven mathematically under special conditions. This is done with the help of the so-called Casimir Effect, named after the physicist Hendrick Casimir. Two electrically uncharged metal plates standing parallel to each other in a vacuum exhibit a force, the Casimir force, which can only be explained by quantum physics (cf.: Müller, 2016, p. 69 f.).Particle-antiparticle pairs fluctuate between the plates in an apparently empty vacuum. They appear together, move away from each other and finally destroy each other, as shown in Fig. 10.

Particle formation   Destruction antiparticle

Figure 10: Particle-antiparticle pairs

The particle-antiparticle pairs cannot be measured directly with a particle detector and are therefore called virtual (cf.: Hawking, 2018, p. 161). Only electromagnetic fields with a frequency that is a multiple of the distance between the plates fit between the metal plates. However, this condition does not apply to the environment outside these plates (cf.: Müller, 2016, p. 70). Thus, there are fewer virtual particles (also called vacuum fluctuations) inside the plates than outside (cf.: Hawking, 2018, p. 161). So the electromagnetic fields outside the plates must compress them with the help of the Casimir force. This quantum effect was proved by the physicist Steve Lamoreaux in 1997 and it followed that this must be negative energy (cf.: Müller, 2016, p. 70).

One can also imagine these particle-antiparticle pairs as a single particle moving through spacetime in a time loop. If such a pair moves into the future, i.e. forward, it is called a particle. If a pair moves towards the past, it is called an antiparticle moving into the future. Stephen Hawking postulated that black holes are not really black and emit radiation and particles. These emissions are virtual pairs of particles and antiparticles. If this pair now comes close to a black hole, one of the pair's partners falls into it. The other partner could escape the immense gravitational force, as illustrated in Fig. 11, and look to an outside, distant observer as if the black hole has emitted the particle (cf.: Hawking, 2018, p. 210).

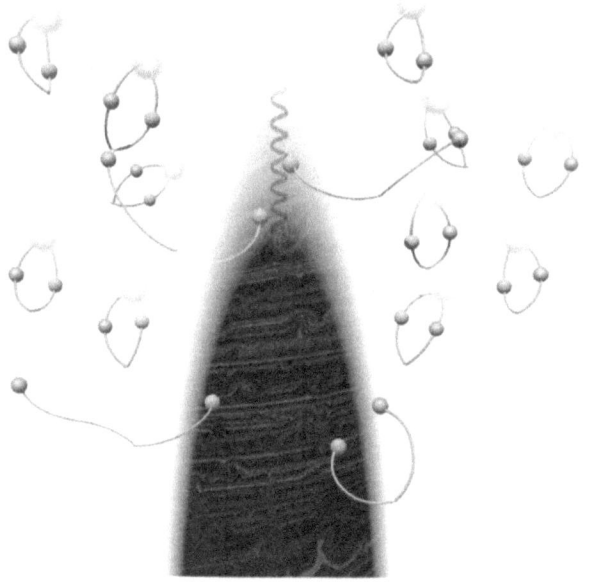

Figure 11: Particle-antiparticle pairs next to a black hole

However, it is also legitimate to claim that the partner that fell into the black hole travels backwards in time and thus moves out of the black hole. At the original point of origin of the virtual particle-antiparticle pair, gravity causes the pair to become a particle that moves forward in time and thus escapes from the black hole. The partner of the pair that falls into the black hole could also be seen as an antiparticle that moves forward in time and thus escapes from the black hole. From this it follows that black hole radiation is an indication that quantum theory allows movements into the past at the microscopic level and that these movements are observable. (cf.: Hawking, 2018 p. 211).

There are theoretically two ways to detect wormholes. One is to look for the wormholes already mentioned quantum mechanical way, where wormholes are brought out of the space-time foam or in a classical way. In the classical way, space-time would be deformed like plasticine without cracks forming. However, it is only possible to avoid the formation of cracks if time is also distorted or shaped. If this were to happen, one would have to be able to move forward and backward in time simultaneously. For a short period of time, it should therefore be possible to transport future events into the past - the principle of a time machine would thus be fulfilled (cf.: Vaas, 2013, p. 178). If one chooses this path, however, a big problem is hidden behind it, namely a so-called naked singularity. This open edge is created in the time between the cutting of space-time and the rejoining of the two. This deformation of spacetime can be compared to a ring formed from a plasticine ball (cf.: Vaas, 2013, p. 178f). Creating a wormhole would be somewhat easier by quantum mechanical means, since no time loops or singularities would be created. However, this requires a theory of quantum gravity, a combination of the GTR and quantum theory (cf.: Vaas, 2013, p. 179).

As already mentioned, tiny wormholes on the Planck scale could exist in quantum foam. Now, if you want to enlarge such a small wormhole and keep it stable, you need the so-called inflation. The inflation in the cosmic area means an immensely fast expansion of space in an extremely short time, which has also increased our universe fractions of a second after the Big Bang. One could use this method, but the problem would be to stop this inflation again (cf.: Vaas, 2013, p. 177 ff.).

To detect a wormhole, one could also look for exotic matter. It would also influence the light in the same way as massive objects, such as black holes, influence the light and thus become observable. When a bundle of light rays falls into a wormhole, it first "contracts" and then expands again after exiting the tunnel. A wormhole therefore has the same effect as a diverging lens. The opposite is true for black holes, which act like converging lenses. Just like black holes, wormholes - caused by the negative mass at the throat - produce a gravitational lensing effect. This phenomenon was calculated by Igor Novikov. It would result in the fact that - in contrast to light deflected by normal mass - there is not just one single increase in the brightness of stars. So-called caustics form, which means that the brightness of a star passing a wormhole increases, decreases and increases again. (cf.: Vaas, 2013, p. 182).

6.3.3 Time travel with the help of wormholes

Imagine two beams of light. One doesn't fly through

a wormhole but takes the longer path around it. The second light beam takes the shortcut through the wormhole. This light beam will reach its destination faster and when it arrives it will observe the light beam that flew around the wormhole, even if it started earlier. So if you were traveling with the light beam that went through the wormhole, you could see into the past when it exits. Under certain circumstances one could even watch oneself how one had prepared the journey (cf.: Müller, 2016, p. 64).

In addition, wormholes could bend a so-called world line of a passing spaceship, i.e. the path of the spaceship through spacetime, so much that it would be bent to a place in time and space where it had already been before. This loop that is created is called a closed time-like curve or time loop. Time-like stands for the orbit, also known as a geodesic, which has a particle of matter in the theory of relativity (cf.: Müller, 2016, p. 65).

Closed time-like curves were first calculated in 1949 by the Austrian mathematician Kurt Gödel. He found a solution to Einstein's field equations which is known today as the Gödel solution. As already mentioned, the solutions of the field equations are spacetimes. In his solution Gödel describes a space-time in which the universe rotates (cf.: Müller, 2016, p. 67 f.). However, Gödel's universe does not expand, so his solution had to be rejected. But it still brings new insights into the physics of relativity. Thus, the GTR theoretically allows to travel through closed time-like curves into the past (cf.: Müller, 2016, p. 68).

Figure 12 shows the idea of "taking" the end of a wormhole on a journey through space and time. While one end is on Earth, the spaceship takes the other on its journey. When returning, less time has passed at the opening that the spaceship took with it than at the opening on Earth. So when you enter the wormhole opening on Earth you could come out at an earlier time at the opening in the spaceship (cf.: Hawking, 2016, p. 145).

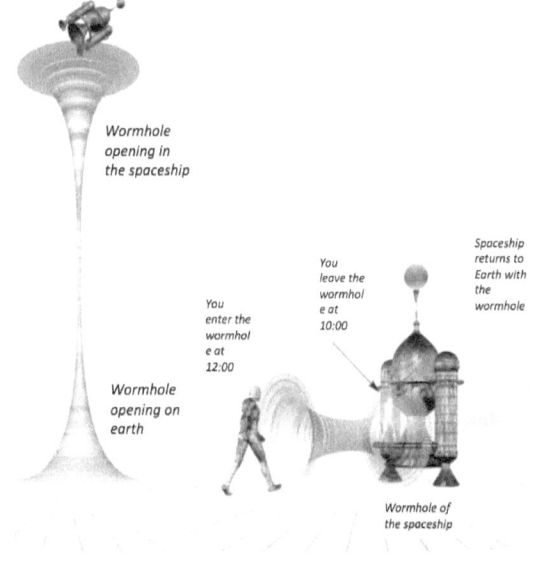

Figure 12: The concept of wormholes

6.3.4 Evidence of a wormhole?

Wormholes have not yet been empirically proven, but there is an extraordinary phenomenon that appears in the constellation of Leo and could possibly be an indication of a wormhole. It is a gravitational lensing effect that Adam Bolton of the University of Hawaii and other researchers discovered and recorded with the Hubble Space Telescope. They found a double ring as shown in Fig. 13.

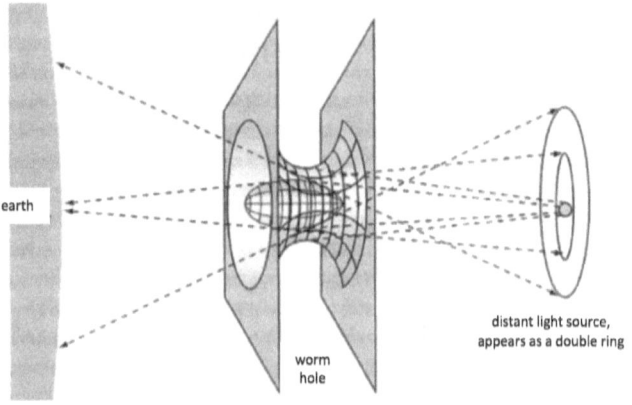

Figure 13: Gravitational lens effect - double ring

This phenomenon was interpreted by scientists as a double Einstein ring. An Einstein ring is created when a gravitational lens is placed exactly in front of a light source and the light, in our view, is fanned out like a ring. This can now be interpreted as a simple double-lens effect, but Pedro Gónzalez-Díaz from the Institute of Fundamental Physics in Madrid believes that this is rather a ring hole, which would be a certain type of wormhole (cf.: Vaas, 2013, p. 184).

7. THE CONCEPT OF THE SUM OVER ALL STORIES

The American physicist Richard Feynman put forward the hypothesis that a particle has several stories. It moves on many different orbits through the space-time continuum as shown in Fig. 14.

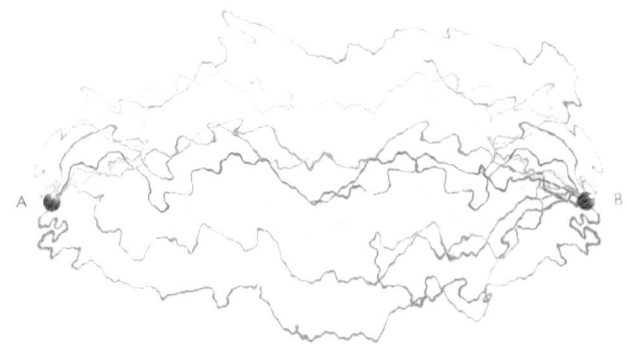

Figure 14: The concept of the sum over all stories

Feynman calculated the probability for a particle to move from A to B by adding up all the waves whose paths lead from A to B. In everyday life, his concept of the "sum of all the stories" is also used, namely it ensures that in macroscopic objects all the orbits cancel each other out but one is left over. Thus, we have the feeling that an event moves from A to B on one orbit (cf.: Hawking, 2016, p. 91).

The sum of all stories, which are spacetimes, of a particle has to be calculated. This includes those that are not compatible with Einstein's equations, i.e. also spacetimes that have a curvature that would be suitable for time travel into the past. Such stories would be, for example, of particles that move in a closed time-like curve, travel faster than light, or move back in time. It turns out that, on the microscopic scale, time travel occurs everywhere. (cf.: Hawking, 2016, p. 155 f.).

8. TIME PARADOXES

Time travel is very controversial among physicists, because some paradoxes can arise when traveling through time. If everyone would own a time machine like today's cars, a real "time travel chaos" would arise. Historically important events could be changed and thus steer the course of history in a different direction (cf.: Kaku, 2017, p. 370 f.)

8.1 The Grandfather Paradox

If you would fly through a wormhole with a spaceship, you could come back before you set off on your journey (cf.: Hawking, 2016, p. 143). Now it would be possible, since you have travelled into the past, to destroy your own spaceship before it starts its journey. This is the so-called grandfather paradox. The idea behind the paradox is that a time traveler murders his grandfather or grandmother before the parents of the time traveler are in the world (cf.: Hawking, 2016, p. 144).

8.2 The twin paradox

Probably the most famous paradox in connection with time travel is the twin paradox. One person A and one person B are twins. B climbs into a space capsule on the twins' 20th birthday and flies, for example, to the star Vega, which is 26 light years away from Earth. The space capsule moves at about 99% of the speed of light. When person B reaches the star, he or she turns around and flies back to Earth. A calculation shows that 52.5 years have now passed on Earth, while for B the entire journey has taken only 7.4 years due to time dilation. So the twin A that remained on earth is 72.5 years old when B returns, while B is only 27.4 years old. B covered the distance in 7.4 years. Mathematically, this means that the space capsule would have had to fly at 7 times the speed of light. According to the theory of relativity, however, the speed of light must not be reached or exceeded. So the journey for B must have been different from the view from Earth (cf.: Putz/ Jahn, 2019, p. 216).

8.3 Paradoxes in general

If you travel into the past, you might also encounter the problem of running into yourself. No one knows what would happen in such a meeting. In the time travel film "Back to the Future II" the girlfriend of the protagonist Marty meets her older self. This meeting has no serious consequences for the course of time or the universe. However, it could certainly cause confusion if there are suddenly many egos in different times. If a time traveler travels into the past, then into the present and again into the past, two of the time traveler's egos would have to exist in that past. If, on the other hand, he travels into the future, it is rather unlikely that he will meet his aged ego, since he has set out on the journey in the present and therefore cannot actually exist in the future (cf.: Müller, 2016, p. 82).

With time travel, a violation of the principle of causality could also occur. The principle of causality describes the fixed sequence of effect and cause. An effect always has a cause first. However, if time loops exist in the universe, effect and cause could occur simultaneously or a contradiction, namely a reversal, of the causality principle could occur. This is why some scientists believe that the laws of nature prohibit closed time-like curves, which would mean that time travel into the past would most likely not be possible (cf.: Müller, 2016, p. 83).

The question why no one has yet encountered a time traveler from the future raises further paradoxes. It could be that time travel - especially into the past - is an impossibility, that it is generally prohibited by the laws of nature, or that it has already happened, but we have not noticed it. A theory of quantum gravity - the most famous candidate is string theory - would provide information as to whether closed time-like curves of the GTR exist and and thus time travel into the past could be made possible at the very least. (cf.: Müller, 2016, p. 85f.).

Stephen Hawking is of the opinion that time travel into the past is inherently forbidden and in 1992 postulated his time protection presumption that would avoid chaos in the universe caused by time travel (cf.: Müller, 2016, p. 86 f.). Stephen Hawking said: "The laws of physics conspire to prevent time travel on the macroscopic scale." (cf.: Hawking, 2018, p. 166). Igor Novikov's consistency conditions could prevent a time traveler from changing history. Everything would be determined and history would be safe from time travelers. Thus one could also understand why we have never been visited by time travelers from the future (cf.: Müller, 2016, p. 87).

8.4 Solution of paradoxes through world lines

With the help of so-called world lines, derived from Einstein's theory of relativity, such paradoxes could possibly be resolved. A world line would be, for example, if you decide not to get up at seven o'clock in the morning and go to work, but to stay in bed for another four hours. Even with this kind of doing nothing, you draw a world line. You do not move in space, but time passes anyway. The world line can be imagined as a vertical line (cf.: Kaku, 2017, p. 375). If you now go to work at eleven o'clock, the world line runs diagonally upwards to the right. World lines have no beginning, no end and cannot break apart, because when you are born you get a mixture of the world lines of your parents. Even if a person dies, the molecules of the body continue to draw their world line (cf.: Kaku, 2017, p. 376). If travel through time is possible, the world line of a time traveler would bend into a closed time-like curve (cf.: Kaku, 2017, p. 379). According to the world line theory, a journey into the past, in which one prevents, for example, one's parents from meeting, would not simply disappear, as for example shown in the film "Back to the Future". Because if you disappear yourself, your own world line disappears as well, which is impossible according to Einstein's theory of relativity. So according to this theory you cannot change the past (cf.: Kaku, 2017, p. 378).

8.5 The Many-worlds-Theory

In order not to change the entire history when travelling back in time, the Many-worlds-Theory would also be a suitable concept. During time travel, the universe is split into two universes, thus creating a parallel universe as shown in Fig. 15. This approach might be in line with quantum theory (cf.: Kaku, 2016, p. 287). The Many Worlds Theory is also presented in the film "Back to the Future". The scientist "Doc Brown" draws a horizontal line on a blackboard, which should represent the time in our universe. Then he draws another line, which has its origin in our universe, but branches off into a parallel universe, which forms when there is a change in the past. So if you - like in the grandfather paradox - murder your parents in the past before you are born, this only means that you have murdered two people who are genetically identical to your parents but not the true parents, because you are in a parallel universe in the past and not in the personally true past. (cf.: Kaku, 2016, p. 288).

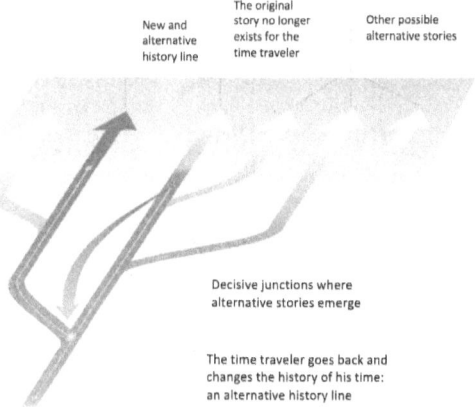

New and alternative history line

The original story no longer exists for the time traveler

Other possible alternative stories

Decisive junctions where alternative stories emerge

The time traveler goes back and changes the history of his time: an alternative history line

Figure 15: Possible solution of time travel paradoxes

The Many-worlds-Theory, which was formulated in 1957 by the physicist Hugh Everett, states that the universe splits up into more and more universes during its evolution. There should therefore be an infinite number of universes that are connected with each other (cf.: Kaku, 2017, p. 413). Stephen Hawking has developed a theory in which wormholes could be used to travel back and forth between universes (cf.: Kaku, 2017, p. 416).

9. CONCLUSION

Whether time travel in the sense of the blockbuster "Back to the future", i.e. a person sits down in a machine that is, for example, a car, moves at the speed of light and arrives at a point in time in the past or the future, ever possible cannot be clearly stated. Scientists are highly divided on this topic. However, humans and other macroscopic objects cannot travel through time with the current state of technology.

Only astronauts who spend a long time in orbit go on a small time journey into the future, because they move at higher speeds and thus the time dilation of the STR begins. They age by a fraction of a second less than the people on Earth (cf.: Vaas, 2013, p. 266). In any case, the GTR would theoretically allow time travel. For example, it allows wormholes and time-like closed curves with which one could travel through time (cf.: Müller, 2016, p.85).

There are no limits to the imagination when building time machines. A time machine could function in such a way that a person could fly in a space capsule at almost the speed of light. Due to the time dilation from the STR, time in the spaceship would pass more slowly than on Earth. The person in the cap sule would thus travel into the future (cf.: Müller, 2016, p. 26). Another way to travel through time would be to "park" a space capsule near a very massive object in space. This time machine is based on the principle of GTR, because time passes more slowly near larger masses, such as black holes. It would thus be possible to travel into the future, because time would pass more slowly in the space capsule than on Earth (cf.: Müller, 2016, p. 49). The probably most popular and at the same time most controversial time and space machines among physicists are wormholes. So far, they are objects that exist only on paper and could connect long distances in space. If one flies through the opening of a wormhole, one could possibly watch oneself preparing for the journey when leaving the wormhole. One could thus look into the past (cf.: Müller, 2016, p. 64). All these methods to travel through time, however, raise many problems and questions that could only be clarified with the help of a "quantum theory of gravity" (cf.: Müller, 2016, p. 86).

When travelling through time, some time paradoxes can arise. For one thing, the grandfather paradox, which mainly refers to the past, could have particularly dangerous effects. Indeed, if people travel back in time, they could murder their grand-parents before their parents were born and thus prevent their own birth (cf.: Hawking, 2016, p. 144). Another paradox is the twin paradox. Here, one twin travels into space and the other remains on earth. When one of them returns from outer space, the latter has aged less than its twin and has thus travelled into the future (cf.: Müller, 2016, p. 24). The possible violation of the principle of causality also speaks against time travel. The order in which events take place could get mixed up (cf.: Müller, 2016, p. 83).

If everyone were allowed to travel through time, under certain circumstances chaos or even war-like conflicts would arise. However, one might not even have to worry about possible chaos scenarios, since nature could, with the help of the chronology protection thesis, maintain the order of events and make time travel impossible (cf.: Hawking, 2016, p. 158). Another solution to these problems, in relation to the past, would be the Many-worlds Theory. Here, every time someone travels into the past, a parallel universe would open up, so to speak, and one would not murder someone in the "true" universe, but in another, "false" parallel universe (cf.: Kaku, 2017, p. 413).

As one can read from the findings, it is very difficult to say whether time travel is at some point possible for humans or not. In the 19th century, however, one would never have believed that it would ever be possible to send a message from Europe to America in less than a second. Therefore, at the moment, impossibilities such as time travel could be reality in a few hundred years. A "quantum theory of gravitation" is needed - a theory that combines GTR and quantum theory - to really be able to say whether time travel can at some point be made a reality for humans as well, and not just in the minds of science fiction fans.

THE REASON WHY

Thank you for reading *Lisa Visintainer's* book! One of our missions at Ilias Thiesseas is to give young authors like *Lisa Visintainer* a voice. Often, junior writers do not get the chance to publish their works, because they are considered too young and inexperienced.

Many people don't know how much work goes into getting our ideas out there. We made the experience that the time to write the actual content takes about **25 percent** of the whole publishing process. We also realised that one simple thing hindered new authors from getting their book out there: experience.

We, at Ilias Thiesseas, also fought hard and made some sacrifices to realise our dream: creating a publishing company which benefits both writers and society. We put our blood and souls into it, and as years went by, we got one yet simple thing: experience.

It was the lack of experience that usually made young authors give up their dreams not the lack of great ideas. They gave up their dreams of presenting society with innovative ideas, of being a writer, of changing the world.

Therefore, we created a new branch in our company that supports new authors. *VWA Publikation Professional Publishing Program*™ is an increasingly popular program for first-time writers to get their ideas into the world. We want authors to focus on what's important to them: writing. Ilias Thiesseas deals with the publishing and also contributes to junior writers' financial independence, so they can fulfil their dreams.

Yes, it took a lot of effort. But after seeing writers come into our offices smiling, we knew that it was worth the months of intense hard work!

We, at Ilias Thiesseas, only have a limited amount of resources and would, therefore, really appreciate your support by leaving a review on this book. Your words really make a difference in keeping dreams of new authors alive.

Thank you for your support!

REFERENCES

Printed media:

Apolin, Martin: Big Bang Physik 8. 1st Edition, Wien: ÖBV, 2008.

Hawking, Stephen: Das Universum in der Nussschale. 7th Edition. München: dtv, 2016.

Hawking, Stephen: Die illustrierte kurze Geschichte der Zeit. 14th Edition. Reinbek bei Hamburg: Rowohlt, 2019.

Hawking, Stephen: Eine kurze Geschichte der Zeit. 23rd Edition. Reinbek bei Hamburg: Rowohlt, 2018.

Hawking, Stephen: Kurze Antworten auf große Fragen. 6th Edition. Stuttgart: Klett-Cotta, 2018.

Jaros, Albert/ Nussbaumer, Alfred/ Nussbaumer, Peter/ Kunze, Hansjörg: Physik compact. Basiswissen 8. 1st Edition. Wien: öbv & hpt, 2007.

Kaku, Michio: Die Physik der unsichtbaren Dimensionen. Eine Reise durch Zeittunnel und Paralleluniversen. 4th Edition. Reinbek bei Hamburg:

Rowohlt, 2017

Kaku, Michio: Die Physik des Unmöglichen. Beamer, Phaser, Zeitmaschinen. 6th Edition. Reinbek bei Hamburg: Rowohlt, 2016.

Müller, Andreas: Zeitreisen und Zeitmaschinen. Heute Morgen war ich noch gestern. 1st Edition, Berlin, Heidelberg: Springer, 2016.

Putz, Bruno/ Jahn, Brigitte: Faszination Physik 7 bis 8. 1st Edition, Linz Veritas, 2019.

Vaas, Rüdiger: Tunnel durch Raum und Zeit. Von Einstein zu Hawking – Schwarze Löcher, Zeitreisen und Überlichtgeschwindigkeit. 6th updated Edition. Stuttgart: Franckh-Kosmos, 2013.

Online Media:

Gast, Robert: Ins Herz der Finsternis. https://www.spektrum.de/magazin/ins-herz-der-finsternis-das-erste-bild-eines-schwarzen-lochs/1647844 [Accessed: 06.09.2019]

Hummel, Philipp: Knackpunkt ist der Charakter des Schwarzen Lochs. https://www.zeit.de/wissen/2014-11/interstellar-physik [Accessed: 06.02.2020]

Pössel, Markus: Träge und schwere Masse. https://www.einstein-online.info/spotlight/traegeschwere/ [Accessed: 05.02.2020]

LIST OF FIGURES

Figure 5: Müller, Andreas: Doppler-Effekt bei relativistischen Geschwindigkeiten [Doppler effect at relativistic velocities], Zeitreisen und Zeitmaschinen. Heute Morgen war ich noch gestern. 1th Edition, Berlin, Heidelberg: Springer, 2016, p.42

Figure 6: Vaas, Rüdiger: Die erste Zeichnung eines Wurmloch [The first drawing of a wormhole], Tunnel durch Raum und Zeit. Von Einstein zu Hawking – Schwarze Löcher, Zeitreisen und Überlichtgeschwindigkeit. 6th aktualisierte Edition. Stuttgart: Franckh-Kosmos, 2013, p. 168

Figure 7: Hawking, Stephen: Strings und deren Weltflächen [Strings and their world surface], Die illustrierte kurze Geschichte der Zeit. 14th Edition. Reinbek bei Hamburg: Rowohlt, 2019, p. 216

Figure 8: Hawking, Stephen: Positive und negative Krümmung von Raum [Positive and negative curvature of surfaces], Die illustrierte kurze Geschichte der Zeit. 14th Edition. Reinbek bei Hamburg: Rowohlt, 2019, p. 203

Figure 9: Hawking, Stephen: Einstein-Rosen-Brücke [Einstein-Rosen Bridge], Die illustrierte kurze

Geschichte der Zeit. 14th Edition. Reinbek bei Hamburg: Rowohlt, 2019, p. 204

Figure 10: Hawking, Stephen: Teilchen-Antiteilchen-Paare [Particle-antiparticle pairs], Das Universum in der Nussschale. 7th Edition. München: dtv, 2016, p. 153

Figure 11: Hawking, Stephen: Teilchen-Antiteilchen-Paare bei einem Schwarzen Loch [Paricle-antiparticle pairs next to a black hole], Das Universum in der Nussschale. 7th Edition. München: dtv, 2016, p. 153

Figure 12: Hawking, Stephen: Das Prinzip von Wurmlöchern [The concept of wormholes], Das Universum in der Nussschale. 7th Edition. München: dtv, 2016, p. 145

Figure 13: Vaas, Rüdiger: Gravitationslinseneffekt – Doppelring, Tunnel durch Raum und Zeit [Gravitational lens effect – double ring]. Von Einstein zu Hawking – Schwarze Löcher, Zeitreisen und Überlichtgeschwindigkeit. 6th aktualisierte Edition. Stuttgart: Franckh-Kosmos, 2013, p. 185

Figure 14: Hawking, Stephen: Die Aufsummierung von Möglichkeiten [The concept of the sum over all stories], Die illustrierte kurze Geschichte der Zeit. 14th Edition. Reinbek bei Hamburg: Rowohlt, 2019, p. 80

Figure 15: Hawking, Stephen: Mögliche Lösung von Zeitreise-Paradoxien [Possible solution of time travel paradoxes], Die illustrierte kurze Geschichte der Zeit. 14th Edition. Reinbek bei Hamburg: Rowohlt, 2019, p. 208